Algebra Fundamentals for Ultrasound Techs

A Student's Guide

Y. S. Eastwood

iUniverse, Inc.
Bloomington

Algebra Fundamentals for Ultrasound Techs
A Student's Guide

iUniverse books may be ordered through booksellers or by contacting:

iUniverse
1663 Liberty Drive
Bloomington, IN 47403
www.iuniverse.com
1-800-Authors (1-800-288-4677)

ISBN: 978-1-4759-7610-6 (sc)
ISBN: 978-1-4759-7609-0 (e)

Library of Congress Control Number: 2013902860

Printed in the United States of America

iUniverse rev. date: 03/14/2013

Table of Contents

Introduction

How often I have looked upon the faces of my students as they
entered my class. I can see that when it comes to math,
a Veil of Mystification is telling from their eyes.
More often than not, they are fearful that they will never
learn what they believe everybody else understands about math.
And so...they keep silent and it is a deafening silence!
And so...my first task, as a teacher, is to lift that
Veil of Mystification so that the deafening silence no longer exists.
When I was asked to write this book, I wrote it for students.
I wrote it for you!

This book is a study of the fundamentals of basic Algebra
as a foundation for the study of physics as it relates to ultrasound
but this little book can be used to study Algebra in any regard.

So how does this little book work? You might be asking yourself this
question. The language of math in this book has been translated
into language that you can understand. There are 33 sections.

This book was written in a progressive fashion so that it is like using
"building blocks." In other words, each section is built upon the
section before it.

There are many "Rules" in this little book. These are the rules of
mathematics. Highlight the rules so that you remember them.
Write them down on a piece of paper to reinforce your memory of them.

One of the most helpful aspects of this little book is that after each
section there is a True/False Quiz. This format was selected
because very often students will guess at multiple choice
whereas with True/False questions, the students must
work out the problems. Having a short quiz after each section
reinforces the learning of that section.

A valuable suggestion is to have plenty of paper and a pencils on-hand when using this book, as well as a highlighter. Work out all problems in the book as well as the True/False Quiz questions.

Also included are pictures for those of you that are visual learners which will help you to identify the concepts.

After all of the sections and True/False quizzes, is a final examination. The answers to all quizzes and this Final Examination are in the back of the book. The answers are written out so that you understand how and why that answer was arrived at.

Included is a very handy Glossary which, again, is written in language that anyone can understand. Afterall, what good is a Glossary if you have to open a dictionary to understand the words in the Glossary.

At the end of this little book is the Appendix. Here you will find examples of the application of the math you will learn to the concepts of ultrasound.

Try to remember that we learn best by making mistakes. If this happens, do not panic. Return to the section previous to the one that you are studying and try the quiz again.

Being a math teacher for many years, every concept that was learned about how to teach math has been incorporated into this book. Had it not been for the many students I have been honored to teach, this book would otherwise not have been written.

So enjoy, have fun, and relax. The Cavalry...help...is on the way!

Y. S. Eastwood

Basic Operations

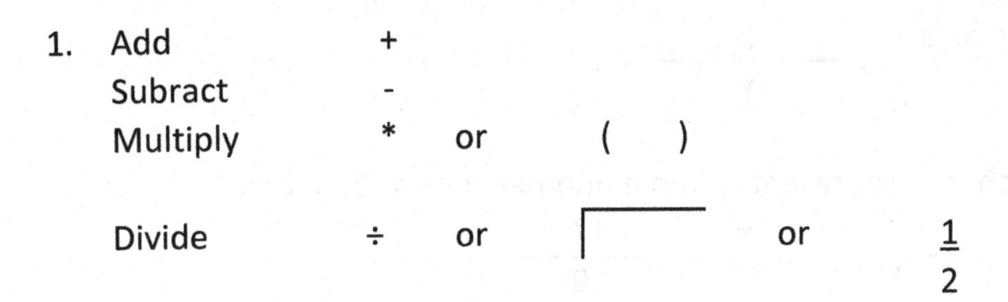

Look at MULTIPLICATION...Why the switch from "X" to "*".......we use the "X" as a variable in Algebra....NOT as a symbol for multiplication!

An Easy Approach to Variables, Constants, and Units

1. A variable is a quantity that can be represented by letters such as x, y, z, and can even be used in fractions: $\frac{x}{y}$

2. A constant is a number which may represent a whole number such as 3, 5, or 17, or even a fraction such as $\frac{3}{5}$ or $\frac{17}{19}$

3. A unit is a measurement such as seconds (sec), microseconds (μsec), MegaHertz (MHz), or even their reciprocals of: MHz or its reciprocal $\frac{1}{\text{MHz}}$

4. Who was Hertz? Why is he important to you? Look at your CELLPHONE!

(1857 - 1894)
Heinrich Hertz

Heinrich Hertz was a German physicist. He discovered that electricity can be transmitted in electromagnetic waves. These waves travel at the speed of light. This made wireless telegraph and radio waves possible.
A unit of frequency of a radio wave is one cycle per second.
So this unit of frequency was named after him. We write it as Hz.
In Ultrasound, the transmission of frequency is expressed as MegaHertz or MHz.

And....if Hertz had not proven this....Well....you would have NO CELLPHONE! And kids would not be texting all the time. And the English language would not be corrupted. Really...What's up with text message of L8r?

An Easy Approach to Variables, Constants, and Units

PRACTICE PROBLEMS

		TRUE	FALSE
1.	X is a variable.	______	______
2.	$\frac{8}{17}$ is a constant.	______	______
3.	A unit of measurement is bunny rabbits.	______	______
4.	Meters is a unit of measurement.	______	______
5.	6 is a variable.	______	______

Fractions, Decimals, and Percentages Finally Understood

MEMORIZE THE FOLLOWING! You should be able to convert fractions to decimals.

FRACTIONS								
$\frac{1}{2}$	$\frac{1}{3}$	$\frac{1}{4}$	$\frac{1}{5}$	$\frac{1}{6}$	$\frac{1}{7}$	$\frac{1}{8}$	$\frac{1}{9}$	$\frac{1}{10}$
NUMERATOR = N								
N → 50	33	25	20	167	14	125	111	10
D → 100	100	100	100	1000	100	1000	1000	100
DENOMINATOR = D								
DECIMALS								
0.5	0.33	0.25	0.2	0.1667	0.14	0.125	0.111	0.1
PERCENTAGES								
50%	33%	25%	20%	16.70%	14%	12.50%	11%	10%

OR......Do it the long way!

Example: $\frac{1}{2} = $

```
    .5
  2|1.0
  - 10
  ----
     0
```

$= \frac{50}{100} = .50 = 50\%$

Convert: $\frac{1}{2} \times \frac{50}{50} = \frac{50}{100} = .50 = 50\%$

1. RULE: $0.01 = \frac{01}{100} = \frac{1}{100} = 1\%$

↓

WHEREAS: $0.1 = \frac{10}{100} = \frac{10}{100} = 10\%$

Let's do a conversion!

If the transmission frequency is 4 MHz (4 Mega Hertz), what is the period?

$$\frac{1}{4MHz} = \frac{1}{4} * \frac{1}{M} * \frac{1}{Hz}$$

$$= \frac{25}{100} * \frac{1}{M} * \frac{1}{Hz}$$

$$= 0.25 * \frac{1}{M} * \frac{1}{Hz}$$

$= .25\mu sec$ (Don't worry about μsec now!)

> You will learn about the meaning of these and others later on.
> M = Mega
> Hz = Hertz
> μ = millionth

PRACTICE PROBLEMS

									TRUE	FALSE
1.	$\frac{1}{2}$	=	$\frac{50}{100}$	=	0.5	=	50%		______	_____
2.	$\frac{1}{5}$	=	$\frac{25}{100}$	=	0.2	=	20%		______	_____
3.	$\frac{1}{6}$	=	$\frac{167}{1000}$	=	0.33	=	16.70%		______	_____
4.	$\frac{1}{9}$	=	$\frac{111}{1000}$	=	0.111	=	11%		______	_____
5.	$\frac{1}{10}$	=	$\frac{10}{100}$	=	0.1	=	10%		______	_____

Reciprocals are a Matter of Flipping...What Fun!

The reciprocal of:

1. x is $\frac{1}{x}$ 4 is $\frac{1}{4}$ sec is $\frac{1}{\text{sec}}$

X can be any quantity. It is an unknown number.

2. RULE: When a number is multiplied by its reciprocal, the product...the answer...is 1.
3. RULE: A fraction is division.
4. RULE: Any number, any variable, any constant, or any unit divided by itself is equal to 1.

$$\frac{1}{x} * \frac{x}{1} = \frac{x * 1}{1 * x} = \frac{x * 1}{x * 1} = 1$$

5. RULE: Reciprocals apply to UNITS also.

Frequency (cycles per second) = Hertz = $\frac{1}{\text{sec}}$

A "flip" means that you turn the fraction upside down.

6. RULE: You must be able to do an EVEN number of flips. An ODD number of flips will not work! ***Watch the arrows below as the fractions "flip!"***

But what if the UNIT is in the denominator? Let's see what happens, shall we?

$$\frac{1}{\text{Hertz}} = \frac{1}{\frac{1}{\text{sec}}}$$

Hertz = $\frac{1}{\text{sec}}$

First: The reciprocal of $\frac{1}{\text{sec}}$ is $\frac{\text{sec}}{1}$

$$\frac{1 * \frac{\text{sec}}{1}}{\frac{1}{\text{sec}} * \frac{\text{sec}}{1}} = \frac{1 * \frac{\text{sec}}{1}}{\frac{1}{\text{sec}} * \frac{\text{sec}}{1}} = \frac{1 * \frac{\text{sec}}{1}}{\frac{\text{sec}}{\text{sec}}} = \frac{1 \text{ sec}}{1} =$$

Answer 1 sec

Another way to do this and why this happens.....

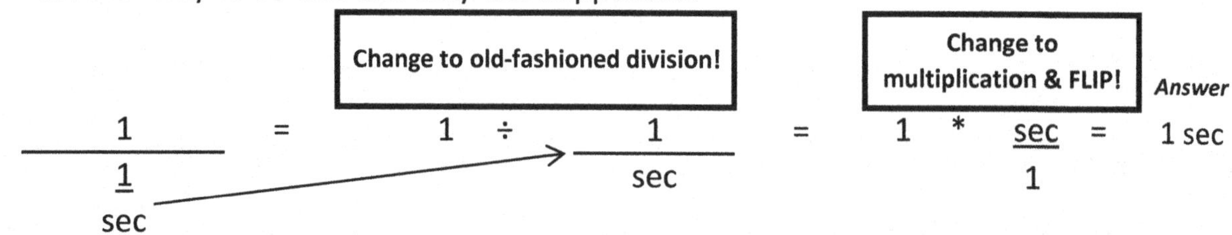

$$\frac{1}{\frac{1}{\text{sec}}} = 1 \div \frac{1}{\text{sec}} = 1 * \frac{\text{sec}}{1} = 1 \text{ sec}$$

Reciprocals are a Matter of Flipping...What Fun! (continued)

7. But what if.......

$$\frac{1}{\frac{1}{\frac{2}{x}}} = \frac{1}{\frac{x}{2}} = \frac{2}{x}$$

$\frac{2}{x}$ Reciprocal is $\frac{x}{2}$

$\frac{x}{2}$ Reciprocal is $\frac{2}{x}$

Answer $\frac{2}{x}$

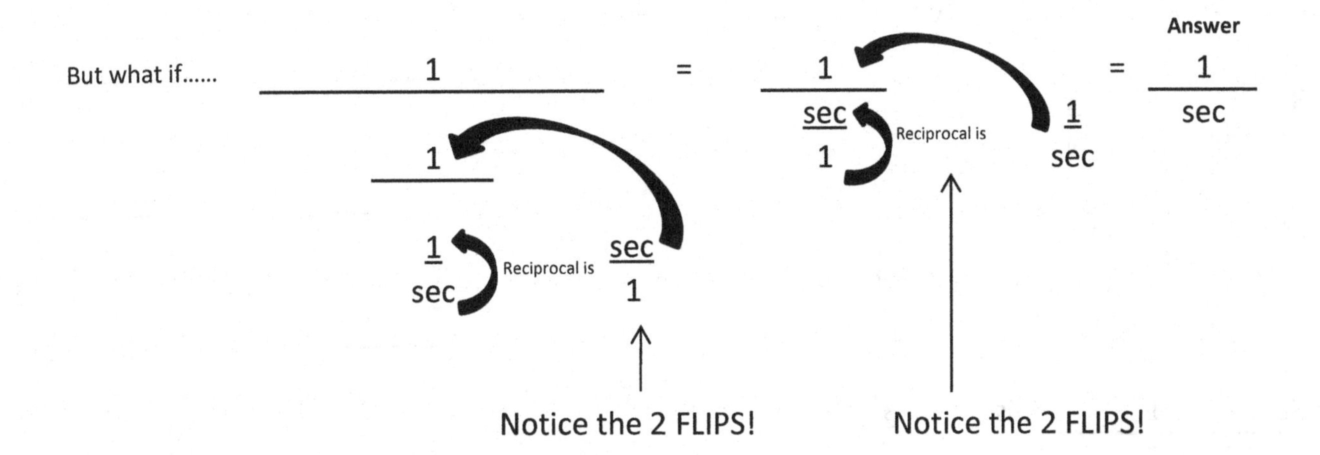

8. RULE: You must be able to do an EVEN number of flips.
An ODD number of flips will not do!

9. Hertz means cycles per second.

$$\text{Hertz} = \text{Hz} = \frac{1}{\text{sec}}$$

SO.......

$$\frac{1}{\text{Hz}} = \frac{1}{\frac{1}{\text{sec}}}$$

Reciprocals are a Matter of Flipping...What Fun!

PRACTICE PROBLEMS

The RECIPROCAL OF:

		TRUE	FALSE
1.	8 is $\frac{1}{8}$	______	______
2.	t is $\frac{1}{y}$	______	______
3.	$\frac{1}{y} = 1$	______	______
4.	Hz is $\frac{1}{\text{sec}}$	______	______
5.	$\dfrac{1}{\dfrac{1}{\frac{3}{X}}}$ is $\frac{3}{x}$	______	______

Exponents Make Multiplication Easy

1. 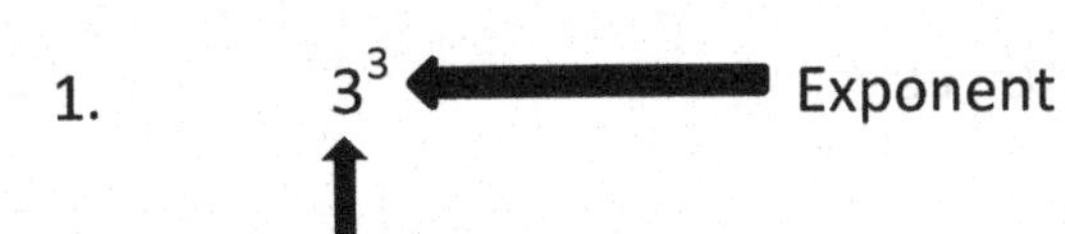

Base

2. RULE: Exponents are a shorthand for multiplication.

3. Think MULTIPLICATION. In the above example, the exponent of 3 is letting you know how many times the number 3 is to be multiplied by itself.
How many times can you multiply 3 by itself?
How many times can you use 3 as a factor?

Answer: The exponent says 3 times. THUS: 3 * 3 * 3 = 27

Look at it this way. It is easier to write 2^5 than 2 * 2 * 2 * 2 * 2. Both equal 32.

4. This same RULE applies to fractions, so......

$$\frac{1}{3}^{3} = \frac{1}{3} * \frac{1}{3} * \frac{1}{3} = \frac{1}{27}$$

Negative Exponents are Positive Reciprocals...Huh???

1. RULE: A number with a negative exponent can be written as a positive reciprocal with a positive exponent.

THUS: $3^{-3} = \frac{1}{3^3} = \frac{1}{27}$

Look at the following example of variables. The same RULES apply.

$$Y^{-D} = \frac{1}{Y^D}$$

THUS:

$$2^{-3} = \left(\frac{1}{2}\right)^3 = \frac{1}{2} * \frac{1}{2} * \frac{1}{2} = \frac{1}{8}$$

LOOK HERE:

$$\left(\frac{1}{2}\right)^{-3} = \frac{2}{1} * \frac{2}{1} * \frac{2}{1} = 8$$

Why did this happen? When the 2 went from being the denominator to being the numerator, the exponent of 3 became positive.

$$\left(\frac{1}{2}\right)^{-3} = \left(\frac{1}{\frac{1}{2}}\right)^3 = 2 * 2 * 2 = 8$$

Whereas:

$$\left(\frac{1}{2}\right)^3 = \frac{1}{2} * \frac{1}{2} * \frac{1}{2} = \frac{1}{8}$$

Adding and Subtracting Positive and Negative Exponents

1. RULE: When multiplying numbers with exponents, add the exponents together!

THUS: $2^2 * 2^3 = 2^{2+3} = 2^5 = 32$

OR EVEN: $Y^A * Y^B = Y^{A+B}$

OR EVEN: $2^4 * 2^{-2} = 2^{4-2} = 2^2 = 4$

WHY? $2^4 * 2^{-2} = \frac{\cancel{2} * \cancel{2} * 2 * 2}{\cancel{2} * \cancel{2}} = 2 * 2 = 4$

Let's do another example

$10^4 * 10^{-2} = \frac{\cancel{10} * \cancel{10} * 10 * 10}{\cancel{10} * \cancel{10}} = 10 * 10 = 100$

2.

RULE: Any expression raised to the zero power is equal to 1.

$y^0 = 1$ Why? $3^3 * 3^{-3} = \frac{\cancel{3} * \cancel{3} * \cancel{3}}{\cancel{3} * \cancel{3} * \cancel{3}} = 1*1*1 = 1$

OR...... $3^{3-3} = 3^0 = 1$

Always remember that any number or variable divided by itself is equal to 1.

Exponents

PRACTICE PROBLEMS

		TRUE	FALSE
1.	$4^3 = 64$	______	______
2.	$\frac{1}{2}^{4} = \frac{1}{8}$	______	______
3.	$\frac{1}{2}^{-4} = 16$	______	______
4.	$4^6 * 4^{-3} = 64$	______	______
5.	$3^6 * 3^{-6} = 1$	______	______

Scientific Notation

1. Scientific Notation is a method for writing very large and very small numbers, using EXPONENTS.

I could give you the age-old piece of information that the sun is 93,000,000 miles away but that is very BORING. Let's look at something more interesting. Any money saved at 5% interest will double every 14 years. If I had lived 2,000 years ago, and opened a savings account with $1 at 5% interest, my savings account TODAY would have.....
$239, 000,000,000,000,000,000,000,000,000,000,000,000,000,000,000,000!
Could you ever imagine doing calculations with this number? Of course not!

THERE ARE 48 OF THE NUMBER 0!

But with SCIENTIFIC NOTATION, you can. It is $2.39 * 10^{50}$!

2. RULE: Write a number as a product of a number ***greater than or equal to 1 and less than 10***, and a power of 10.
   Thus $5.2 * 10^4$ or $5.23 * 10^4$ is OK...........BUT $52.1*10^4$ or $.521*10^4$ is NOT OK!
3. RULE: For large numbers in scientific notation, move the decimal point to the left until there is only ONE WHOLE NUMBER to the left of the decimal point.
   THUS: 52,000 = 5.2 * 10^4
   WHY? 5 2 0 0 0 . = 5.2 * 10^4
   When you move from right to left the exponent is positive.
4. RULE: For very small numbers in scientific notation, move the decimal point to the right until there is only ONE WHOLE NUMBER to the left of the decimal point.
   THUS: 0.00052 = 5.2 * 10^{-4}
   Notice how the exponent of 4 became -4. When you move from left to right, the exponent becomes a negative exponent.
   WHY? 0 . 0 0 0 5 2 = 5.2 * 10^{-4}
5. It is possible to multiply very large numbers and very small numbers by using scientific notation.
   THUS: 4,300,000,000 * 0.000002 = 8,600
   $4.3 * 10^9$ * $2.0 * 10^{-6}$ = 8,600
   WHY?

$$\begin{array}{r} 4.3 \\ * \; 2 \\ \hline 8.6 \end{array} \quad * \quad \begin{array}{r} 10^9 \\ * \; 10^{-6} \\ \hline 10^{9-6} \end{array} = 10^3 = 8.6 * 10^3 = 8,600$$

<u>Scientific Notation</u>

PRACTICE PROBLEMS

						TRUE	FALSE
1.	71,000	=	7.1	*	10^{4}	______	______
2.	0.00000071	=	71.0	*	10^{-5}	______	______
3.	640	=	64	*	10^{2}	______	______
4.	270,000	*	0.002	=	$5.4 * 10^{2}$	______	______
5.	0.0000064	=	6.4	*	10^{-6}	______	______

The Basics of the Metric System

1. RULE: In the United States of America, for medicine and science, we use the metric system that is based on the powers of 10.

2. MEMORIZE THE FOLLOWING TABLE!

Meaning	Prefix	Symbol	Exponent	Number
billion	giga	G	10^{9}	1,000,000,000
million	mega	M	10^{6}	1,000,000
thousand	kilo	k	10^{3}	1,000
hundred	hecto	h	10^{2}	100
ten	deca	da	10^{1}	10
tenth	deci	d	10^{-1}	0.1
hundredth	centi	c	10^{-2}	0.01
thousandth	milli	m	10^{-3}	0.001
millionth	micro	μ	10^{-6}	0.000001
billionth	nano	n	10^{-9}	0.000000001

3. Most often, we will use G, M, k, m, and μ.

4. NOTE: Notice how the table is divided in half so that the positive exponents are exact reciprocals of the negative exponents! Notice how some of the prefixes are very, very similar.

5. And REMEMBER the RULE: Any expression raised to the power of zero is equal to 1.

Reciprocals and the Metric System

1. There is a inverse (opposite) relationship between Frequency and Period of Time.

$$\text{Period (Time)} = \frac{1}{\text{Frequency}} \quad \textbf{OR} \quad \text{Frequency} = \frac{1}{\text{Period (Time)}}$$

In the following example, knowing the FREQUENCY, what is the PERIOD or TIME?

FREQUENCY PERIOD or TIME

$$\frac{1}{5MHz} = \frac{1}{5} * \frac{1}{M} * \frac{1}{Hz}$$

$$= .2 * \frac{1}{\frac{1}{\mu}} * \frac{1}{\frac{1}{sec}}$$

NOTE: How did M become $\frac{1}{\mu}$? How did Hz become $\frac{1}{sec}$?

M is 10^6. But when it went below the 1, it became negative. So to make it positive, it was changed to $\frac{1}{\mu}$ Remember that μ is 10^{-6} and it is an exact reciprocal of M which is 10^6.	Remember this from the beginning. Hertz = Hz = $\frac{1}{sec}$ SO....... $\frac{1}{Hz} = \frac{1}{\frac{1}{sec}}$

So...let's go back to our problem, shall we?

$$.2 * \frac{1}{\frac{1}{\mu}} * \frac{1}{\frac{1}{sec}}$$

$.2 * \mu * sec =$ 0.2μsec OR
0.2 microseconds
is the PERIOD or TIME.

Reciprocals and the Metric System (continued)

We know that.......
There is a relationship between Frequency and Period of Time, so that....

$$\text{Period} = \frac{1}{\text{Frequency}} \quad \text{OR} \quad \text{Frequency} = \frac{1}{\text{Period}}$$

6. Let's do the problem that we did before but in REVERSE!
If given the PERIOD, we want to know the FREQUENCY. Ready? Let's DO IT!

$$0.2\ \mu\text{sec} = \frac{1}{?}$$

WAIT A MINUTE....THIS PROBLEM IS NOT THE SAME AS BEFORE....WHAT'S UP WITH THIS?
You'll be able to figure this out! Believe it or not...YOU CAN DO IT!

Period (Time) — Frequency

$$.2 * \mu * \text{sec} = \frac{1}{?} = \frac{2}{10} * \mu * \text{sec} = \frac{1}{?}$$

Let's take the inverses.

No inverse here!	Inverse of:	Remember....M is 10^6 and μ is 10^{-6}. They are INVERSE !	Inverse of:
$\frac{2}{10} = \frac{1}{5}$	μ is $\frac{1}{M}$		sec is $\frac{1}{Hz}$

$$0.2\ \mu\text{sec} = \frac{1 * 1 * 1}{5 * M * Hz}$$

If the Period is.. — The Frequency is...

$$0.2\ \mu\text{sec} = \frac{1}{5MHz}$$

Reciprocals and the Metric System for Frequency and Period

Frequency ➡ Period	Examples
In Ultrasound, the transmission frequency is in MegaHertz, or MHz. Thus typical periods would be in microseconds or μsec. Period and Transmission Frequency are inverses of each other. $\frac{1}{\text{Period}} = \text{Frequency}$ OR $\text{Period} = \frac{1}{\text{Frequency}}$	$\frac{1}{0.2\mu\text{sec}} = 5\text{MHz}$ OR $0.2\mu\text{sec} = \frac{1}{5\text{MHz}}$ Remember that Mega (M) is 10^6, whereas the reciprocal is micro (μ) is 10-6.
Pulse Repetition Frequency, or PRF, is done in kiloHertz, or kHz. Pulse Repetition Period, or PRP, is an inverse of PRF. $\frac{1}{\text{PRP}} = \text{PRF}$ OR $\text{PRP} = \frac{1}{\text{PRF}}$	$\frac{1}{0.2\text{msec}} = 5\text{kHz}$ OR $0.2\text{msec} = \frac{1}{5\text{kHz}}$ Remember that kilo (k) is 10^3, whereas the reciprocal is milli (m) is 10^{-3}.
Frame Rate (frequency measured in Hertz) and Frame Time have a reciprocal relationship. It is $\frac{1}{\text{Frametime}} = \text{Frame Rate}$ OR $\text{Frametime} = \frac{1}{\text{Frame Rate}}$ Frame Time and Frame Rate determine how many images can be transmitted on the screen.	$\frac{1}{0.2\text{sec}} = 5\text{Hz}$ OR $0.2\text{sec} = \frac{1}{5\text{Hz}}$

Reciprocals and the Metric System for Frequency and Period

PRACTICE PROBLEMS

				TRUE	FALSE
1.	300cm	=	3 meters	______	______
2.	$\frac{1}{2\text{MHz}}$	=	0.5 μsec	______	______
3.	90 liters	=	9000ml	______	______
4.	$\frac{1}{3\text{MHz}}$	=	0.33 μsec	______	______
5.	$\frac{1}{4\text{MHz}}$	=	0.11 μsec	______	______

Converting Between Units In The Numerator

Notice the following change of numbers as we change from one unit of measurement to another.

12 donuts = 1 Dozen

60 minutes = 1 Hour

As we change from a smaller unit of measurement to a larger unit of measurement, the number in front of the larger unit of measurement became SMALLER!

1. RULE: In the NUMERATOR, as the unit becomes bigger, the number in front becomes ***smaller.***

2. RULE: In the NUMERATOR, as the unit becomes smaller, the number in front becomes ***bigger.***

3. RULE: The speed of sound through soft tissue is $1540 \frac{\text{meters}}{\text{second}}$ OR $1540 \frac{\text{m}}{\text{sec}}$

4. Example of going from a SMALLER UNIT to a BIGGER UNIT in the NUMERATOR.

If the speed of sound is $1540 \frac{\text{meters}}{\text{second}}$ ....what is it in kilometers?

$$1540 \frac{\text{m}}{\text{sec}} = \_\_\_\_\_ \frac{\text{km}}{\text{sec}}$$

SO...... The unit of measurement, meters, is becoming BIGGER and, therefore, the number in front of kilometers must become SMALLER!

WHY? 1 meter = 0.001 kilometers

REMEMBER FROM OUR METRIC TABLE....... kilo = 10^3 = 1,000

THUS.... $1\ 5\ 4\ 0. \frac{\text{m}}{\text{sec}} = 1.54 \frac{\text{km}}{\text{sec}}$

NOTE: Move the decimal point RIGHT to LEFT 3 places as kilo = 10^3

If you look closely, it makes sense. A kilometer is made up of 1,000 meters. In other words, it takes 1,000 meters to equal a kilometer.

REMEMBER....

60 minutes = 1 Hour

12 donuts = 1 Dozen

Converting Between Units In The Numerator (continued)

5. Example of going from a BIGGER UNIT to a SMALLER UNIT in the NUMERATOR.

If the speed of sound through soft tissue is 1540 $\frac{\text{meters}}{\text{second}}$ ....what is it in millimeters?

$$1540 \frac{\text{m}}{\text{sec}} = \_\_\_\_\_\_ \frac{\text{mm}}{\text{sec}}$$

SO...... The unit of meters is becoming SMALLER and, therefore, the number in front of millimeters must become BIGGER!

WHY? 1 meter = 1000 millimeters

REMEMBER FROM OUR METRIC TABLE....... milli = 10^{-3}

THUS.... 1 5 4 0 . 0 0 0 $\frac{\text{m}}{\text{sec}}$ = 1,540,000 $\frac{\text{mm}}{\text{sec}}$

NOTE: Move the decimal point LEFT to RIGHT three decimal points as milli = 10^{-3}

Converting Between Units In The Denominator

1. RULE: In the DENOMINATOR, as the unit measurement becomes BIGGER, the number in front of the larger unit of measurement becomes BIGGER!
2. RULE: In the DENOMINATOR, as the unit of measurement becomes SMALLER, the number in front of the smaller unit of measurement becomes SMALLER!

3. Example of going from a BIGGER UNIT to a SMALLER UNIT in the DENOMINATOR.

If the speed of sound through soft tissue is.....

$$1540 \ \frac{\text{meters}}{\text{second}}$$

....what is it in microseconds (μsec)?

LOOK AT THIS PROBLEM VERY CLOSELY. The numerator is changing is also.

SO.....There are 2 steps to this problem. We must change the numerator AND the denominator.

$$1540 \ \frac{\text{m}}{\text{sec}} = \underline{\qquad} \ \frac{\text{mm}}{\mu\text{sec}}$$

STEP 1:

First, let's look at the numerator and change from meters (m) to millimeters (mm).
AND...FOR THE MOMENT...Let's not worry about the denominator.
We know that meters (m) stands alone and is not raised to a power.
We know that millimeters (mm) is raised to the power of 10^{-3}

THEREFORE: $$1540 \ \frac{\text{m}}{\text{sec}} = 1{,}540{,}000 \ \frac{\text{mm}}{(?\ \mu\text{sec})}$$

STEP 2:

NOW..... Let's deal with the units in the DENOMINATOR.

NOW...Let's change the denominator back so that $\frac{\text{mm}}{\text{sec}}$ will be $\frac{\text{mm}}{\mu\text{sec}}$

We know that seconds (sec) stands alone and is not raised to a power.
We know that milliseconds (μsec) is raised to the power of 10^{-6}

$$1{,}540{,}000 \ \frac{\text{mm}}{(?\ \mu\text{sec})} = 1\ 5\ 4\ 0\ 0\ 0\ 0. \ \frac{\text{mm}}{(?\ \mu\text{sec})}$$

$$1540 \ \frac{\text{m}}{\text{sec}} = 1.54 \ \frac{\text{mm}}{\mu\text{sec}}$$

Converting Between Units In The Denominator (continued)

SHORTCUT: $1540 \frac{m}{sec} = \_\_\_\_\_\_ \frac{mm}{\mu sec}$ $\frac{\text{millimeters} = 10^{-3}}{\text{microseconds} = 10^{-6}}$

RULE: When dividing with exponents, subract the exponents.

RULE: A negative number minus a negative number equals a positive number.

SO THAT: $\frac{10^{-3}}{10^{-6}} = 10^{-6--3} = 10^{-6+3} = 10^{-3}$

A negative number minus a negative number equals a positive number!

Remember to start with the exponent in the denominator!

SO THAT: $1540 \frac{m}{sec} = 1\ 5\ 4\ 0.\ \frac{mm}{\mu sec}$

SO THAT: The Answer is....... $1.54 \frac{mm}{\mu sec}$

Converting Between Units

PRACTICE PROBLEMS

		TRUE	FALSE
1.	$\frac{1}{2\text{MHz}} = 0.3\ \mu\text{sec}$	______	______
2.	$\frac{1}{2\text{MHz}} = 0.5\ \mu\text{sec}$	______	______
3.	$\frac{1}{6\text{MHz}} = 0.167\ \mu\text{sec}$	______	______
4.	$\frac{1}{3\text{MHz}} = 0.25\ \mu\text{sec}$	______	______
5.	$1200\ \frac{\text{m}}{\text{sec}} = 1.2\ \frac{\text{mm}}{\mu\text{sec}}$	______	______

Equations, Relationships, and Graphing

1. RULE: A RELATIONSHIP involves 2 variables.

2. RULE: When solving problems involving Relationships, there are 2 parts.
   1. What is the direction? Up or down?
   2. What is the amount, what is the factor, by which either goes up or down?

   In other words.....If X changes, what happens to Y ?

DIRECT RELATIONSHIPS $Y \propto X$

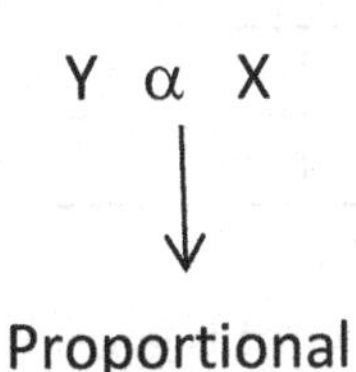

Proportional

What is proportional? Well, let's look at a simple example in your kitchen.
For every cup of rice, you add two cups of water.

X	Cups of Rice	1	2	3
Y	Cups of Water	2	4	6

In other words: $2 \propto 1$

3. RULE: In a DIRECT RELATIONSHIP between variables, when one variable increases the other variable increases.

   RULE: Also, in a DIRECT RELATIONSHP between variables, when one variable decreases, the other variable decreases.

   THUS: In a DIRECT RELATIONSHIP, if X gets BIGGER, then Y gets BIGGER.
   In a DIRECT RELATIONSHIP, if X gets SMALLER then Y gets SMALLER.

4. Example:
   How much money do you spend on hot dogs if each hot dog costs $2.00?
   Let Y = money spent on hotdogs.
   Let X = number of hot dogs purchased.
   WHAT IS THE EQUATION?????
   Money spent on hot dogs = $2.00 multiplied by the number of hot dogs purchased!

   OR

   Y = $2.00 * X

   Notice that $2.00 is a constant....IT DOES NOT CHANGE! The price of hot dogs is fixed.
   $ is the unit....what is being measured.

   So, if you bought 2 hot dogs, how much money did you spend?
   Y = $2.00 * X
   Y = $2.00 * 2
   Y = $4.00

Equations, Relationships, and Graphing (continued)

DIRECT RELATIONSHIPS (continued)

5. In graphing, Y stands for the Y axis; the line that goes up and down.
   X stands for the X axis; the line that goes from side to side.
   Y (How much I spent on hot dogs) = $2.00 * X (How many hot dogs I bought)

OR Y = $2.00 * X

THUS: Y = $2.00 * 2

Y = $4.00

Let's build a table.....

OR

X	0	1	2	3	4	5
Y	0	2	4	6	8	10

RC	X	Y	RC
1	0	0	2
	1	2	
1	2	4	2
	3	6	
1	4	8	2
	5	10	

AS WE CAN SEE.....When X got bigger, Y got bigger.
It is proportional by a factor of 2....$2 This is constant.
The $2 DOES NOT CHANGE!
So when X is 1, Y is 2 and when X is 2, Y is 4 and so forth and so on.

NOW...Let's graph the table!

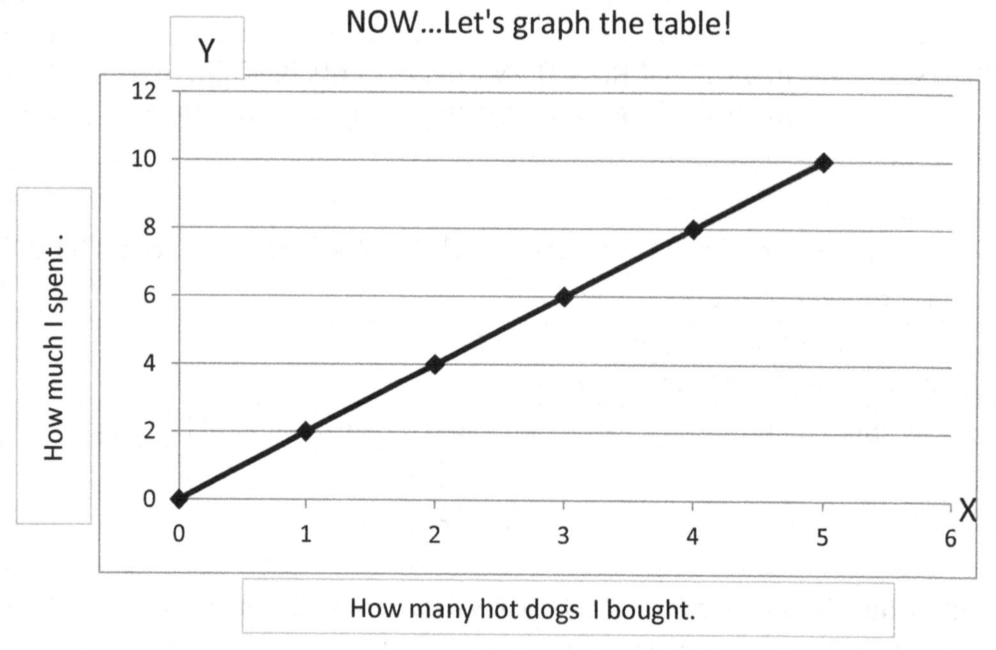

5. RULE: In Linear Relationships, the ***RATE OF CHANGE*** (***RC***), between the variables is the same.

   For Example: Double the time you peddle your bicycle, double the distance.

6. RULE: In a DIRECT RELATIONSHIP, the line that is graphed is a straight line. This is a **LINE**AR RELATIONSHIP.

Equations, Relationships, and Graphing (continued)

INDIRECT (inverse) RELATIONSHIPS

1. RULE: In an INDIRECT (inverse) RELATIONSHIP between variables, when one variable increases the other variable decreases.
2. RULE: Also, in a INDIRECT (inverse) RELATIONSHP between variables, when one variable decreases, the other variable increases.

THUS: In an INDIRECT RELATIONSHIP, if X gets BIGGER, then Y gets SMALLER.
In an INDIRECT RELATIONSHIP, if X gets SMALLER then Y gets BIGGER.

3. Remember this concept? The inverse of $X = \frac{1}{X}$

IF..... $X = Y$ THEN... The Inverse would be $X = \frac{1}{Y}$

Remember: A RELATIONSHIP involves at least 2 variables.

IF...In Direct Relationships, THEN...In Indirect Relationships

SO.... IF......$Y \ \alpha \ \frac{1}{X}$

THEN.... IF..... X goes up by a factor of, say 2, then Y goes down by a factor of 2.

Example: $Y \ \alpha \ \frac{1}{X}$ IF.. $X = 2$ THEN $Y \ \alpha \ \frac{1}{2}$

IF.. $X = 4$ THEN $Y \ \alpha \ \frac{1}{4}$

For Example: Double the speed that you peddle your bicycle, reduce your travel time by one-half.

Equations, Relationships, and Graphing (continued)

INDIRECT (inverse) RELATIONSHIPS (Continued)

Let's build a table.....

$$Y \propto \frac{1}{X}$$

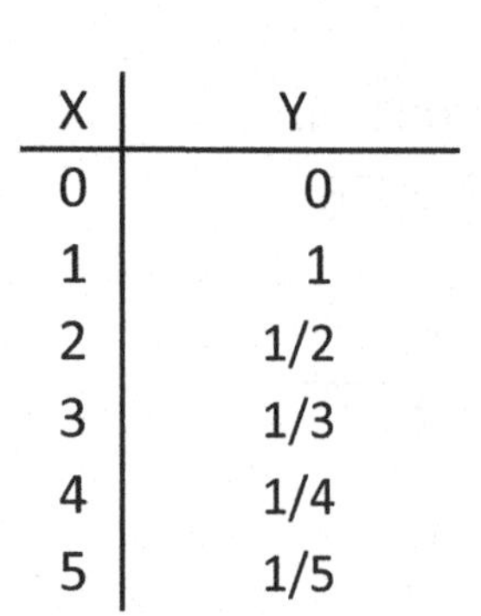

X	Y
0	0
1	1
2	1/2
3	1/3
4	1/4
5	1/5

OR

X	0	1	2	3	4	5
Y	0	1	1/2	1/3	1/4	1/5

NOW...Let's graph the table!

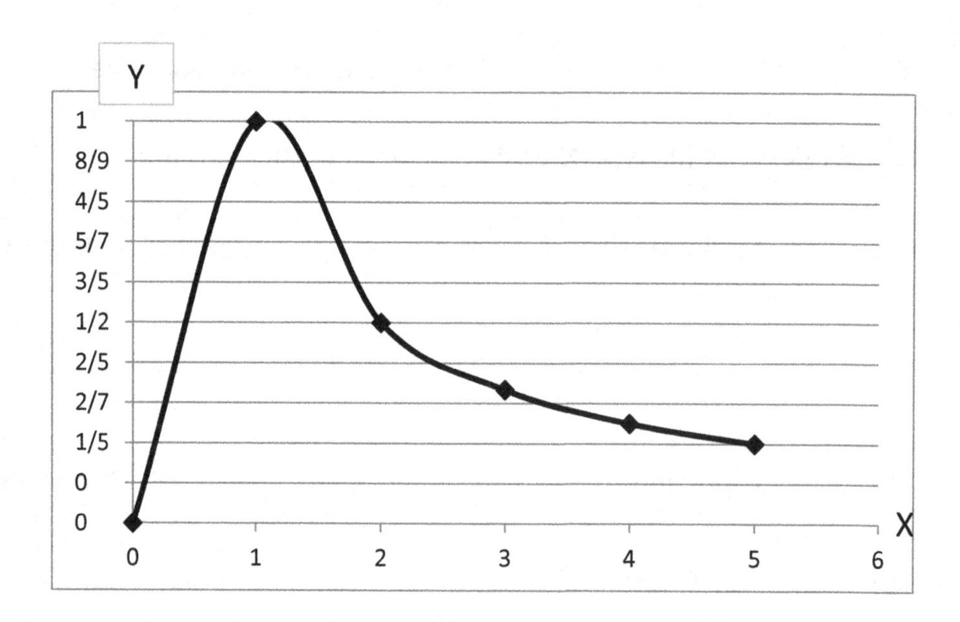

5. RULE: IF.... in graphing a INDIRECT RELATIONSHIP, and if the line that is graphed is not straight and as X gets bigger and Y gets smaller, OR VICE VERSA, then we have a NON-LINEAR, INDIRECT RELATIONSHIP.

6. RULE: **<u>Non</u>**-**<u>line</u>**ar means that when you graph the line, it is NOT straight!

Non-Linear Proportional Relationships Made Easy

1. RULE: A non-linear proportional relationship exists between variables that when one variable increases by a certain amount, the other variable increases by a HUGE amount!

$$Y \quad \alpha \quad X^2$$

If we look at the equation stated above, exactly what does the RULE for non-linear relationships mean? Well, first of all, the words, "non-linear" mean that if we graph it, the line will NOT be straight. Also, as X is getting BIGGER, Y is getting a lot BIGGER!

Let's build a table.....

$Y \quad \alpha \quad X^2$

For Example: If X = 2, then $X^2 = 4$

X	Y
0	0
1	1
2	4
3	9
4	16
5	25

OR

X	0	1	2	3	4	5
Y	0	1	4	9	16	25

NOW...Let's graph the table!

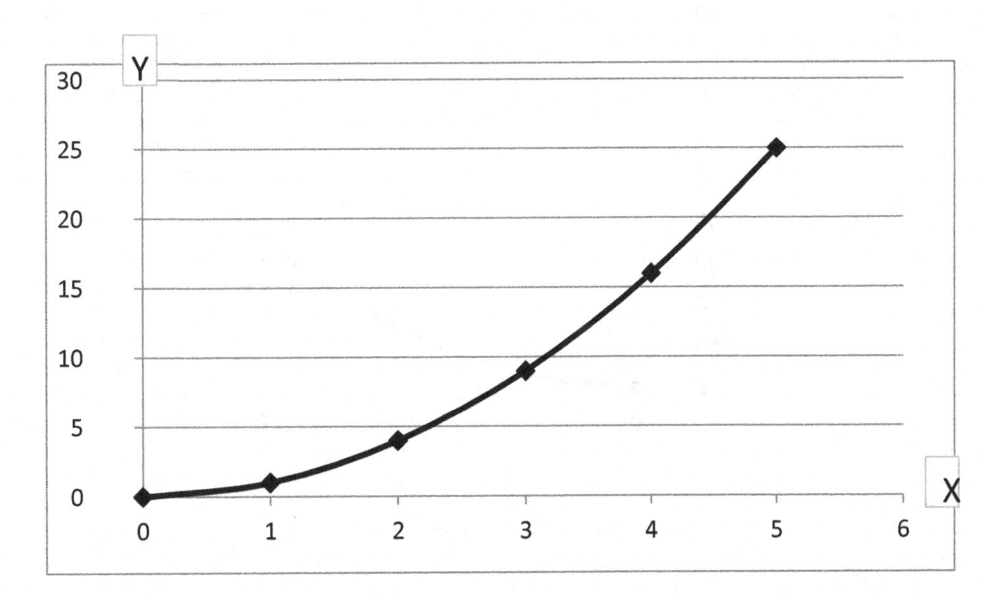

Non-Linear Proportional Relationships Made Easy (continued)

1. Let's do another equation and table!

$Y \ \alpha \ 4X^2$ X^2 is now multiplied by 4

X	Y
0	0
1	4
2	16
3	36
4	64
5	100

OR

X	0	1	2	3	4	5
Y	0	1	16	36	64	100

Let's learn how to graph a Non-Linear Proportional Relationship.

1. If X = 2 then X^2 = X * X or 2 * 2 = 4.
2. What is Y proportional to? It is proportional to X^2 which is 4!
   Now...the last step...See the 4 in front of the X^2?
   What is X^2 again? You remember. It is 4.
3. Multiply X^2 which is 4 by the 4 that is in front of the X^2.
   And your answer is?......16! ***YOU DID IT!***
4. ***Thus, when X is 2, Y is 16.***

NOW...Let's graph the table!

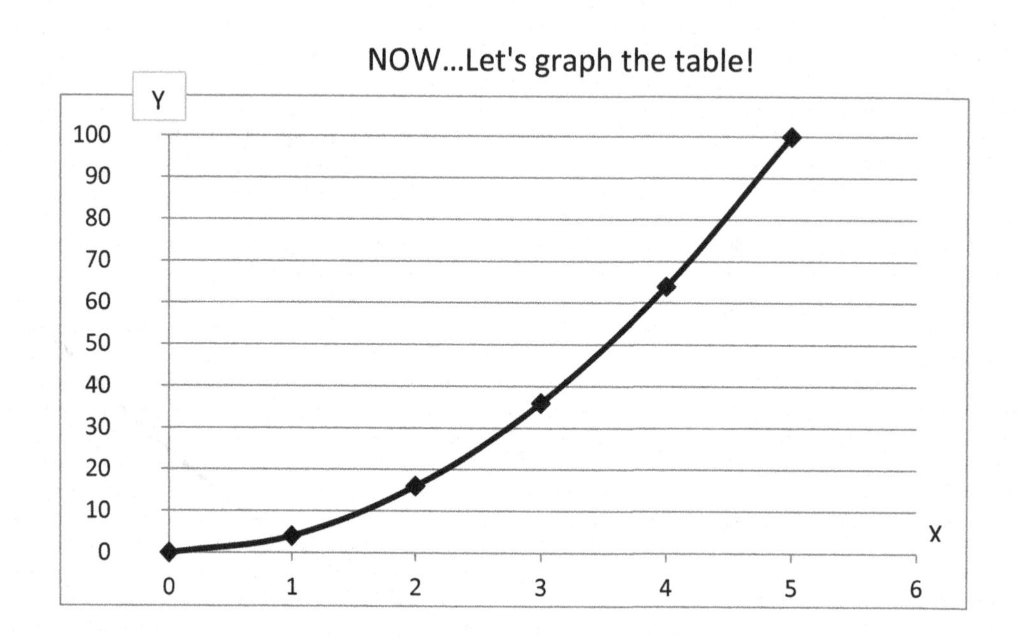

Non-Linear Inverse Proportional Relationships

1. RULE: A non-linear inverse proportional relationship exists between variables when one variable increases by X% and the other variable decreases by a different percent.

$$Y \quad \alpha \quad \frac{3}{X^2}$$

If we look at the equation stated above, exactly what does the RULE for non-linear relationships mean? Well, first of all, the words, "non-linear" means that if we graph it, the line will ***NOT*** be straight. Also, as X is getting BIGGER, Y is getting a lot *smaller* !

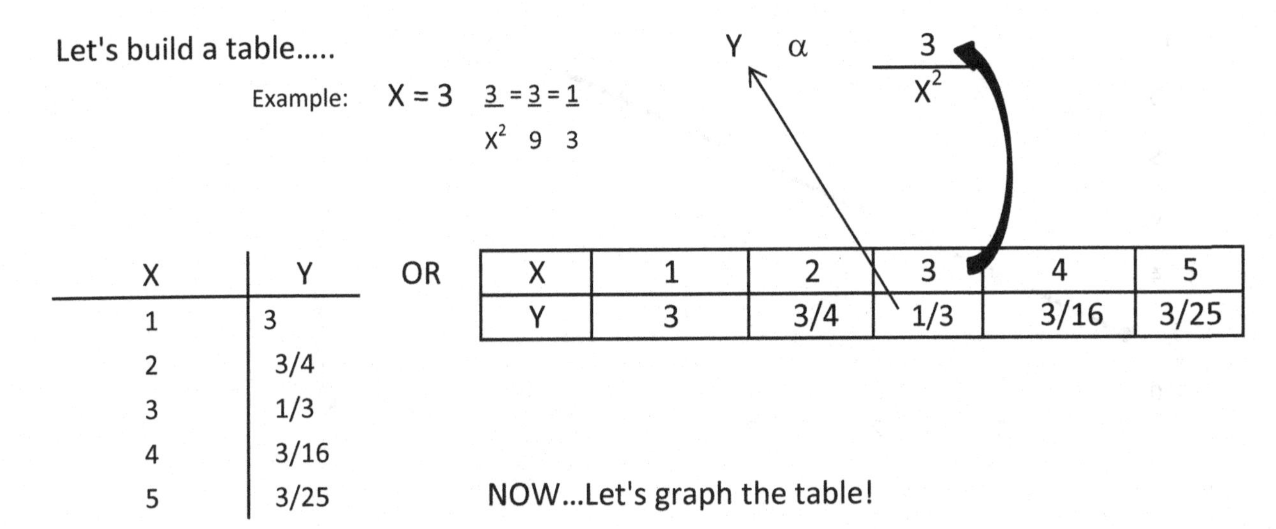

Let's build a table.....

Example: $X = 3 \quad \frac{3}{X^2} = \frac{3}{9} = \frac{1}{3}$

$$Y \quad \alpha \quad \frac{3}{X^2}$$

X	Y
1	3
2	3/4
3	1/3
4	3/16
5	3/25

OR

X	1	2	3	4	5
Y	3	3/4	1/3	3/16	3/25

NOW...Let's graph the table!

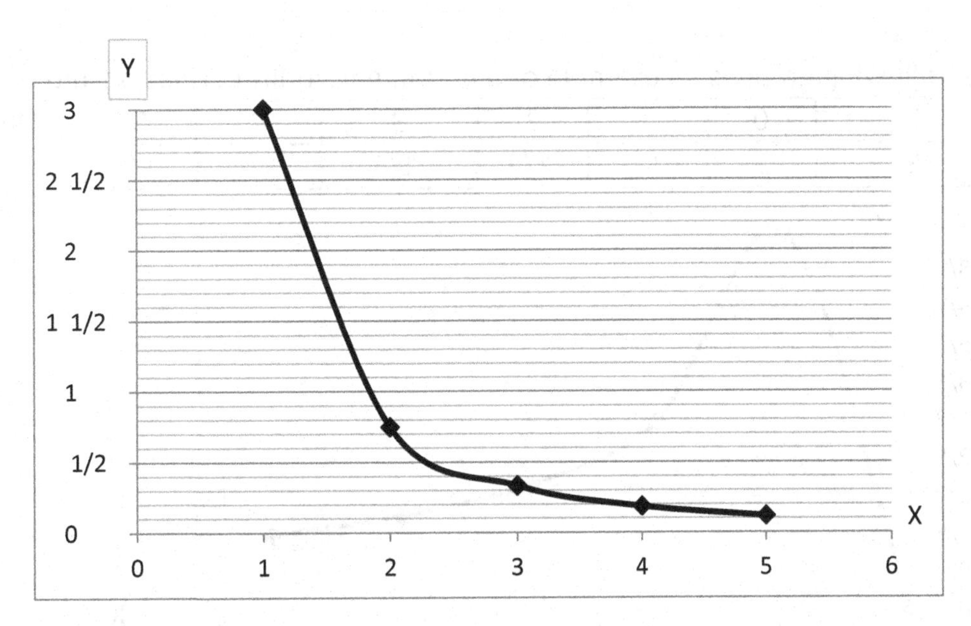

Equations and Relationships

PRACTICE PROBLEMS

1. The following graph shows a direct, linear relationship.

X	0	1	2	3	4	5
Y	2	4	6	8	10	12

TRUE ____ FALSE ____

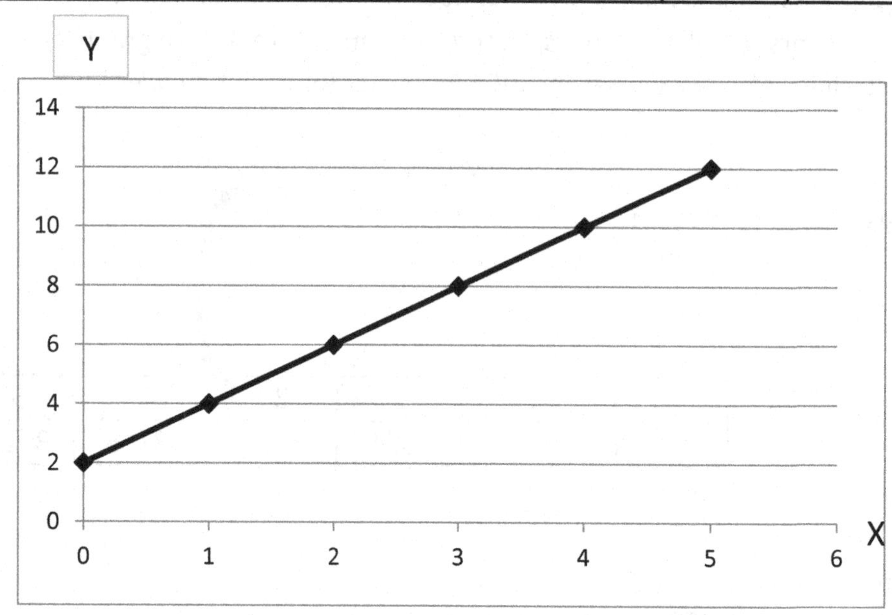

2. The following graph is an example of a non-linear, indirect relationship.

X	0	1	2	3	4	5
Y	0	1	1/2	1/3	1/4	1/5

TRUE ____ FALSE ____

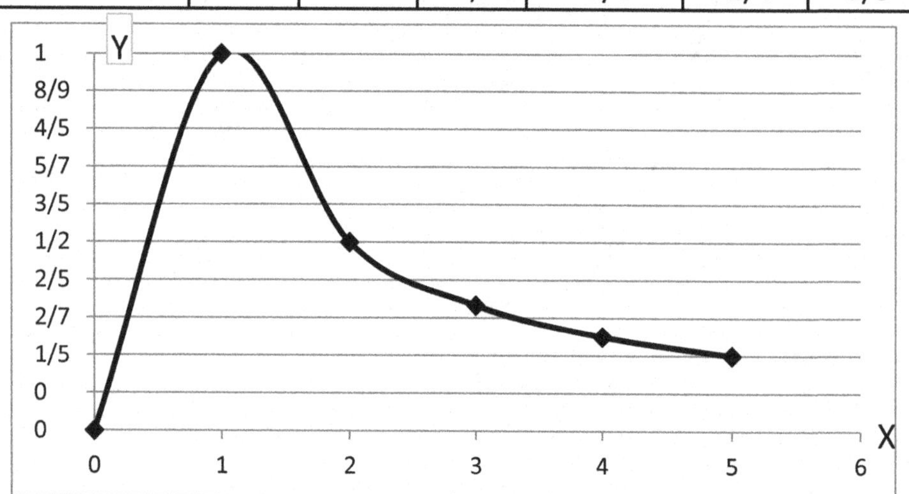

Determining Relationships

$$Y = \frac{a * b^2}{c^3}$$

1. From this equation, we know two things.
   1. If one of the variables, a b or c, change, Y will change.
      For example: If b is equal to 3, then Y increases by 9 because of the exponent of 3. Remember $3^2 = 3 * 3 = 9$.
   2. If one of the variables change, the others will not change.
      a, b, and c are all different. Remember that!
      If a changes, would b change? Would c change? NO!

A Chart to help you Determine Relationships!

Y is proportional to a	
Linear and a Direct Relationship	
If a doubles, y also doubles.	
a	y
2	2
4	4
8	8
16	16

Y is proportional to b^2	
Non-Linear and a Direct Relationship	
If b doubles, y increases by a factor of 2^2 or 4.	
b	y
2	4
4	16
8	64
16	256

$4 * 4 = 16$
$4 * 4 * 4 = 64$
$4 * 4 * 4 * 4 = 256$

Y is α to $\frac{1}{c^3}$	
Non-Linear and an Inverse Relationship	
If c doubles, y decreases by 2^3 or 8.	
c	y
2	8
4	64
8	512
16	4096

$8 * 8 = 64$
$8 * 8 * 8 = 512$
$8 * 8 * 8 * 8 = 4096$

Remember that
c is the denominator
so that $1/2^3$ (1/8)
and $1/4^3$ (1/64) is
larger than $1/8^3$ (1/512).

Determining Relationships

PRACTICE PROBLEMS

Are the following statements directly related, inversely related, or unrelated?

	TRUE	FALSE
1. Cholesterol level and longevity are directly related.	______	______
2. IQ and a person's weight are directly related.	______	______
3. Alcohol intake and sobriety are inversely related.	______	______
4. The size of a tree and it's age are directly related.	______	______
5. Education level and salary are directly related.	______	______

Solving for a Particular Variable in an Equation

REMEMBER....the See Saw when you were a child?

1. RULE: Whatever you do on one side of an equation, you must do to the other so that the equation is balanced, so that it is EQUAL!

For Example: The Distance Equation is......

$$d = r * t$$

d → distance, r → rate, t → time

BUT....What if I wanted to solve for time? What would I do?

$$\frac{d}{r} = \frac{\cancel{r}}{\cancel{r}} * t$$

$$\frac{d}{r} = t$$

The Distance Equation

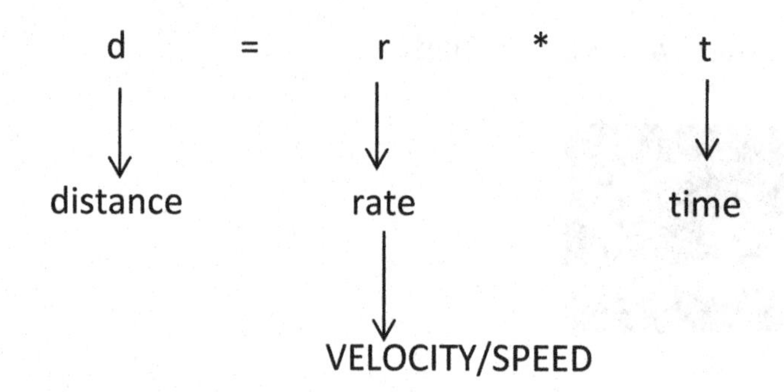

1. The Distance Equation is the one we use for driving, for bicycling, for running and many other things. This is the SAME EQUATION that we use in ultrasound.

2. RULE: In using the Distance Equation in Ultrasound, Ultrasound is a reflective mode and, therefore, in transmitting sound into the body, the sound will reflect back.

3. But what if we wanted to solve for time? Let's do it!

Step 1: $d = r * t$

Step 2: $\frac{d}{r} = \frac{\cancel{r}}{\cancel{r}} * t$

Step 3: $\frac{d}{r} = t$

The Distance Equation (continued)

Let's solve a problem using the Distance Equation. Let's solve for Time.

REMEMBER.....Time is measured in microseconds or μsec.

Let's make the Distance 1 centimeter!

1. $$1\ cm = 1540 \frac{m}{sec} * t$$

2. $$\frac{1\ cm}{1540 \frac{m}{sec}} = \frac{1540 \frac{m}{sec}}{1540 \frac{m}{sec}} * t$$

3. $$\frac{1\ cm}{1540 \frac{m}{sec}} = t$$

3. Let's do a PROBLEM and see how much you have learned!

$$\frac{1\ cm}{1540 \frac{m}{sec}} = t$$

4. Uh oh! We have a problem! If rate is supposed to be in meters and distance is supposed to be in centimeters, and we have to get the $\frac{m}{sec}$ away from there...what do we do???????

Let's make the centimeters into meters so that we can get rid of the meters.

NOTE: Centimeters is 10^{-2}.

$$\frac{1\ cm = 1 * 10^{-2} m}{1540 \frac{m}{sec}} \quad \text{OR} \quad \frac{1 * 10^{-2}\ m}{1540 \frac{m}{sec}} = \frac{.01\ m}{1540 \frac{m}{sec}}$$

$$\frac{.01\ sec}{1540} = 1540\overline{)0.01} \;(0.0000065) = 0.0000065\ sec$$

Uh oh! We have a problem! Time is supposed to be in microseconds(μsec)!

No problem.......μ or micro is 10^{-6}. So..... Time = 0.0000065 = 6.5 μsec

REMEMBER: Ultrasound is a reflective mode so the time will be double!

THUS: 6.5 μsec * 2 = 13 μsec

Time and Distance in Ultrasound Finally Understood

1. Here is a handy chart that you should memorize for time and distance in Ultrasound. This is necessary to know as an ultrasound technologist.

Time	*Distance*
6.5 μsec	1 cm
13 μsec	2 cm
26 μsec	4 cm
39 μsec	6 cm
52 μsec	8 cm
65 μsec	10 cm
78 μsec	12 cm
91 μsec	14 cm
104 μsec	16 cm
117 μsec	18 cm

2. REMEMBER: If the imaging depth is 1 cm, then the distance traveled is 2 cm or 13 μsec.

AND THIS WILL BE ON THE TEST!
Let's do some problems that might be on a test!

1. How much time is required to image a structure at a depth of 5 cm?

2. How much time is required for sound to travel a distance of 10 cm (assuming soft tissue)?

What is the difference in the 2 questions?

1. Well...in the first one, the key word is "image" and, therefore, it is a "roundtrip" excursion in Ultrasound. Thus, the answer is: 6.5 * 2 = 13 μsec

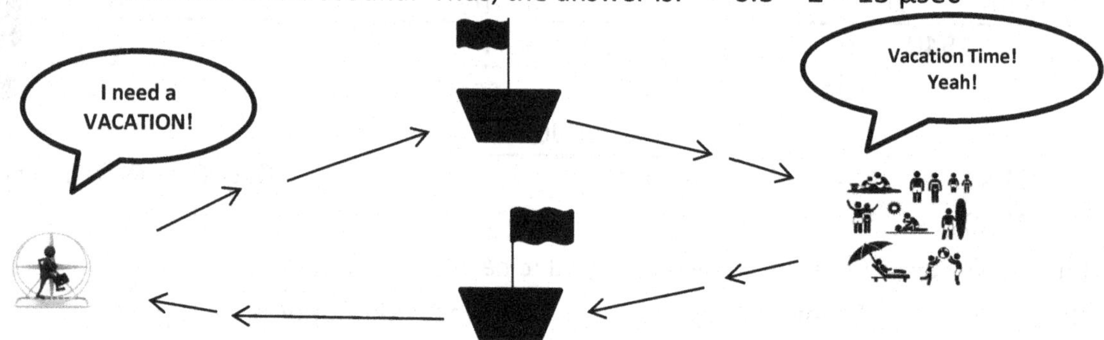

2. In the second question, it is only in one direction. It is a one-way trip.
Answer: 6.5 μsec

FINAL EXAMINATION

1. Which of the following statements about reciprocals is NOT true?
A. If we multiply reciprocals, the product will be 1.
B. Time and frequency are reciprocals.
C. Reciprocals can be applied to constants and variables, but not to units.
D. The reciprocal of transmit period is transmit frequency.

2. 1/10 equals what percent?
A. 0.10%
B. 1%
C. 10%
D. 0.01%

3. 4^3 = ?
A. 48
B. 4
C. 12
D. 64

4. 0.06 MHz is equal to:
A. 6 Hz
B. 60 kHz
C. 0.000006 kHz
D. 6 kHz

5. If the period of a wave is 0.2 microseconds, what is the transmit frequency?
A. 2 Hz
B. 50 MHz
C. 5 MHz
D. 20 Hz

Please CONTINUE to the NEXT PAGE

6. Which of the following is NOT an appropriate unit for volume?
A. cubic miles
B. seconds
C. liters
D. cm^3

7. Which of the following about metric symbols is NOT true?
A. centi = c = 10^{-2}
B. μ = micro = 10^{-6}
C. k = kilo = 10^{-3}
D. M = mega = 10^{6}

8. How much time is required to image to a depth of 3 cm?
A. 78 μsec or microseconds
B. 39 μsec or microseconds
C. 18 μsec or microseconds
D. 117 μsec or microseconds

9. How much time is required is required for sound to travel 8 cm?
A. 52 μsec or microseconds
B. 104 μsec or microseconds
C. 78 μsec or microseconds
D. 10.4 μsec or microseconds

10. In regards to contributions to ultrasound, who was Hertz?
A. He is a guy that has a rental car business.
B. He is one of the minority owners of the NFL Atlanta Falcons.
C. He discovered that electricity can be transmitted in electromagnetic waves.
D. He was a French sociologist. His field of study was the sociology of religion.

PRACTICE PROBLEMS...ANSWER KEY

An Easy Approach to Variables, Constants and Units

Page 3

1. T
2. T
3. F
4. T
5. F

Fractions, Decimals and Percentages Finally Understood

Page 5

1. T
2. F
3. F
4. T
5. T

Reciprocals area a Matter of Flipping ...What Fun!

Page 8

1. T
2. F
3. F
4. T
5. T

Exponents

Page 12

1. T
2. F
3. T
4. T
5. T

Scientific Notation

Page 14

1. T
2. F
3. F
4. F
5. T

Reciprocals and the Metric System for Frequency and Period

Page 19

1. T
2. T
3. F
4. T
5. F

Converting Between Units

Page 24

1. F
2. T
3. T
4. F
5. T

Equations and Relationships

Page 32

1. T
2. T

Determining Relationships

Page 34

1. F
2. F
3. T
4. T
5. T

FINAL EXAMINATION...ANSWER KEY

1. Which of the following statements about reciprocals is NOT true? Answer is: C

A. If we multiply reciprocals, the product will be 1.
B. Time and frequency are reciprocals.
C. Reciprocals can be applied to constants and variables, but not to units.
D. The reciprocal of transmit period is transmit frequency.

2. 1/10 equals what percent?

Answer is: C

A.	0.10%	This is 1/1000 or 10/10,000
B.	1%	This is .01 or 1/100
C.	10%	This is 10/100 or 1/10
D.	0.01%	This is 1/10,000

3. $4^3 = ?$

Answer is: D

A.	48	4 * 3 * 4
B.	4	4 * 1 = 4
C.	12	4 * 3 = 12
D.	64	4^3 = 4 * 4 * 4 = 64

4. 0.06 MHz is equal to:

Answer is: B

A.	6 Hz	No quantifier in front of Hz.
B.	60 kHz	kilo is 10^3. Mega is 10^6. $10^6 - 10^3 = 10^3$. Move 3 decimals to right.
C.	0.00006kHz	Moved in the wrong direction by three decimals.
D.	6 kHz	Only moved decimal over 2 spaces, not by 3.

Answer is: C

5. If the period of a wave is 0.2 microseconds, what is the transmit frequency?

A.	2 Hz	No quantifier in front of Hz.
B.	50 MHz	Too large in conversion. See Answer C
C.	5 MHz	.2 = 2/10 = 1/5
D.	20 Hz	No quantifier in front of Hz. Also, no inverse computated.

Please CONTINUE to the NEXT PAGE

FINAL EXAMINATION...ANSWER KEY (continued)

6. Which of the following is NOT an appropriate unit for volume? Answer is: B

A. cubic miles — Cubic is three dimensional.

B. seconds — A measurement of time, not volume.

C. liters — Liters is three dimensional.

D. cm^3 — The m^3 implies three dimensions.

7. Which of the following about metric symbols is NOT true?

A. centi = c = 10^{-2} Answer is: C

B. μ = micro = 10^{-6}

C. k = kilo = 10^{-3} — k = kilo = 10^3 not 10^{-3}.

D. M = mega = 10^6

8. How much time is required to image to a depth of 3 cm?

A. 78 μsec or microseconds Answer is: B

B. 39 μsec or microseconds — How long does it take to get the image?

C. 18 μsec or microseconds — This is a round trip. 3 cm + 3 cm = 6 cm.

D. 117 μsec or microseconds — 6m = 39 μsec or microseconds.

9. How much time is required is required for sound to travel 8 cm? Answer is: A

A. 52 μsec or microseconds — This question is asking how long it takes to get to what we want to image.

B. 104 μsec or microseconds

C. 78 μsec or microseconds — This is a one-way trip. 8cm = 52 μsec.

D. 10.4 μsec or microseconds

10. In regards to contributions to ultrasound, who was Hertz? Answer is: C

A. He is a guy that has a rental car business.

B. He is one of the minority owners of the NFL Atlanta Falcons.

C. He discovered that electricity can be transmitted in electromagnetic waves.

D. He was a French sociologist. His field of study was the sociology of religion.

APPENDIX A
Application of Math to Ultrasound

Solving for a Particular Variable in an Equation

Let's solve for a variable in a particularly difficult equation!
By the way, this equation is used a great deal in ultrasound.

DOPPLER EQUATION

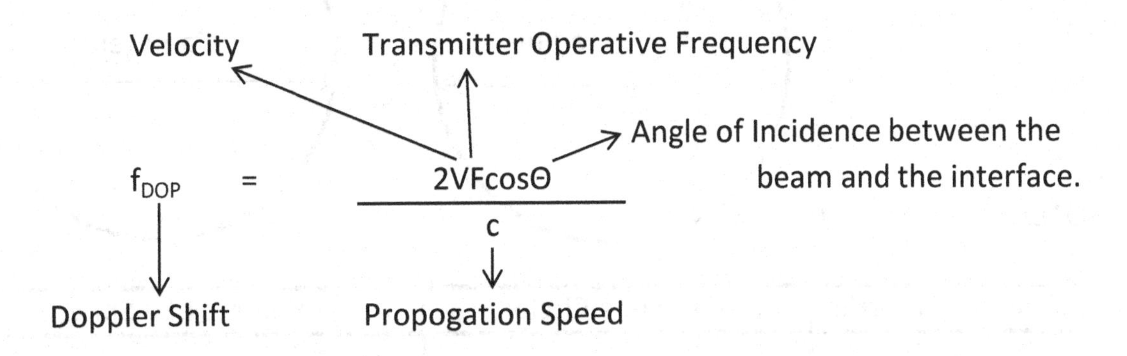

Let's solve for **V**.....Velocity!

$$f_{DOP} = \frac{2 * \mathbf{V} * F * \cos * \Theta}{c}$$

$$f_{DOP} = \frac{2 * F * \cos * \Theta * c * \mathbf{V}}{2 * F * \cos * \Theta * c}$$

$$\frac{f_{DOP}c}{2F\cos\Theta} = \frac{\cancel{2} * \cancel{F} * \cancel{\cos} * \cancel{\Theta} * \cancel{c} * \mathbf{V}}{(\cancel{2} * \cancel{F} * \cancel{\cos} * \cancel{\Theta} *) \cancel{c}}$$

$$\frac{f_{DOP}c}{2F\cos\Theta} = \mathbf{V}$$

Dimensional Measurement of a Circle

1. When it comes to circles...... r = radius d = diameter

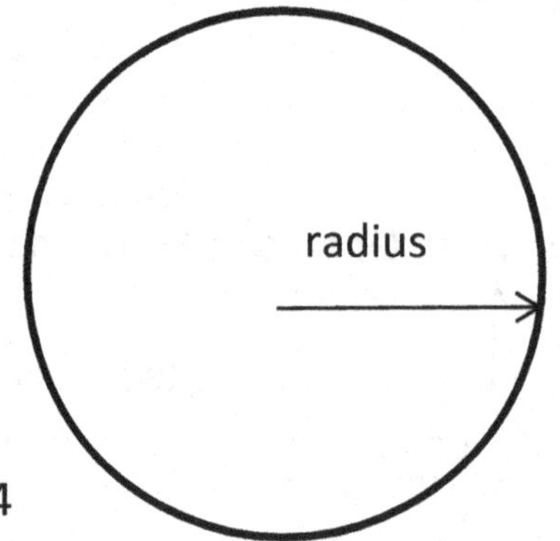

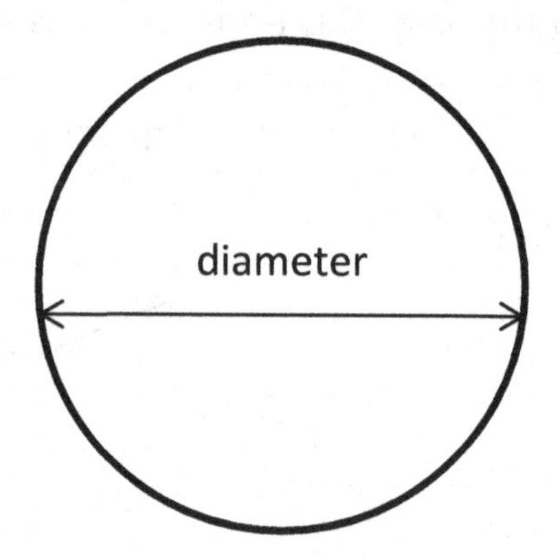

NOTE: $\pi = 3.14$

Circumference of a Circle	Area of a Circle	Volume of a Sphere
Circumference is the distance around the circle.	Area is all that is within the circle.	Volume is related to a sphere. Think of a ball! And...how much it takes to fill up the ball.
Formula: $2 \pi r$ It is one dimensional because $r = r^1$.	Formula: πr^2 It is two dimensional because $r^2 = r * r$.	Formula: $4/3 \pi r^3$ It is two dimensional because $r^3 = r * r * r$.
Units expressed in: meters If r = 2, then circumference = C = 2 * 3.14 * 2 C = 12.56 m	Units expressed in: $meters^2$ If r = 3, then area = $A = 3.14 * 3^2$ A = 3.14 * 9 $A = 28.2 m^2$	Units expressed in: $meters^3$ If r = 4, then volume = $V = 4/3 * 3.14 * 4^3$ V = 4/3 * 3.14 * 64 $V = 267.95 m^3$
Proportion Circumference α r	Proportion Area α r^2	Proportion Volume α r^3
		Volume can be expressed in meters, liters, feet but not in time such as seconds.

Rules for Percent Change in a Linear Relationship

The following pages are absolutely necessary to know if you are going to become an ultrasound technologist.

Let's look at the following problem....

If the radius of a vessel is reduced by 30%, what is the percent residual area?

What does this question mean?
It means that if a vessel goes down by 30%, how big is it now in percent?
And, NO, the answer is not 70%

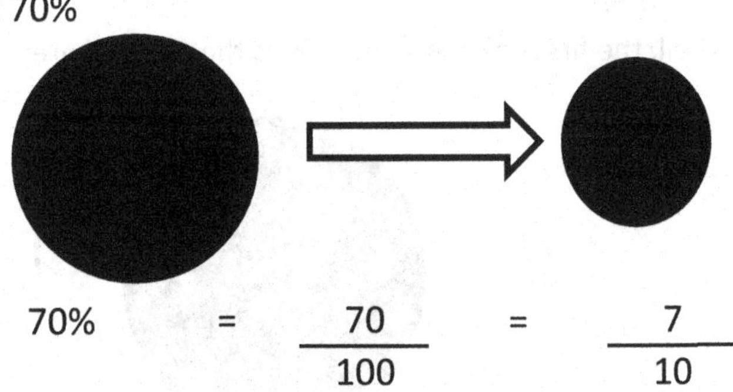

Follow these steps to solve this problem.

1. Change the percent to a fraction.

First, let's subract....100% - 30% = 70%
Now, let's make it into a fraction.

$$70\% = \frac{70}{100} = \frac{7}{10}$$

SO THAT.............

$$r_{new} = \frac{7}{10} r_{old}$$

2. Write a mathematical relationship.
In this case, the problem stated, "....percent residual AREA."

$$\text{Area} \quad \alpha \quad r^2$$

$$\text{Area}_{new} \quad \alpha \quad r^2_{new}$$

3. Apply the relationship to the fraction.

$$\text{Area}_{new} \quad \alpha \quad r^2_{new} = \left(\frac{7}{10}\right)^2 r_{old} = \frac{7 * 7}{10 * 10} = \frac{49}{100}$$

4. Convert the fraction back to a percent.

$$\frac{49}{100} =$$ 49% of the original or the old or RESIDUAL area.
In other words...what is left!

Rules for Percent Change in a Linear Relationship (continued)

Let's look at the previous problem on Page 30.

If the radius of a vessel is reduced by 30%, what is the percent residual area?
Answer: 49% of the original or the old or RESIDUAL area.

But supposed we were asked...
If the radius of a vessel is reduced by 30% what is the percent decrease in the area?

What is the difference between the two questions?

Well, the first one is asking what is the residual area, or what is the area that is remaining.

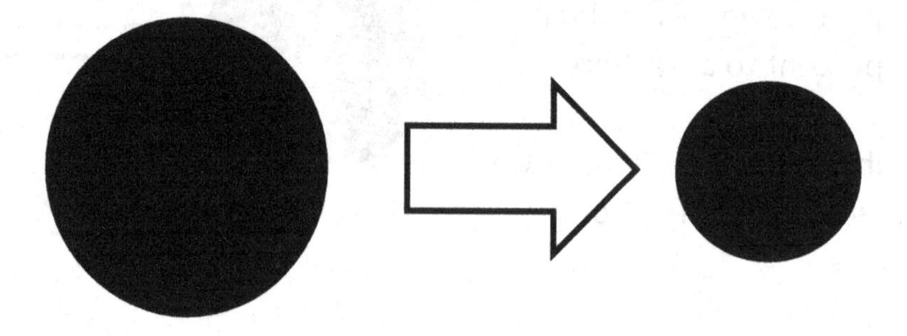

The second one is asking how much the area decreased.

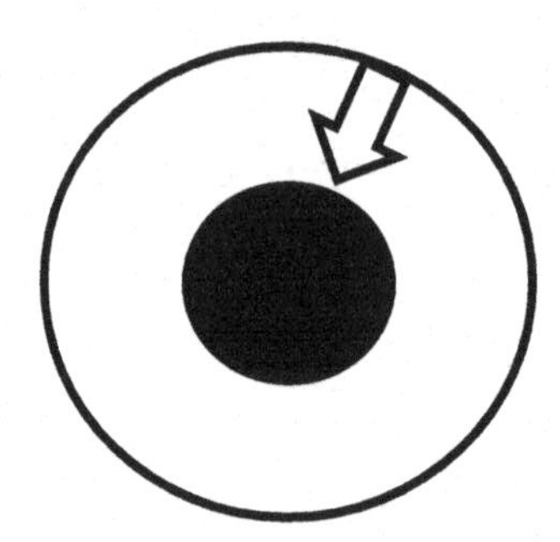

How do we solve this? Simple!

100%	
-49%	Residual area (what is left).
51%	Decrease in area.

Rules for Percent Change in a Non-Linear Relationship

Let's look at the following problems....

1. If the residual vessel radius is 30%, what is the percent residual area?
2. If the residual vessel radius is 30%, what is the percent decrease in the area?

What do these questions mean?

In the first question, it is asking what is the area of a vessel if the radius is NOW 30%

In the second question, it is asking how much the area decreased by.
This will involve subtraction.

Radius was 100%.

Radius is now 30%.

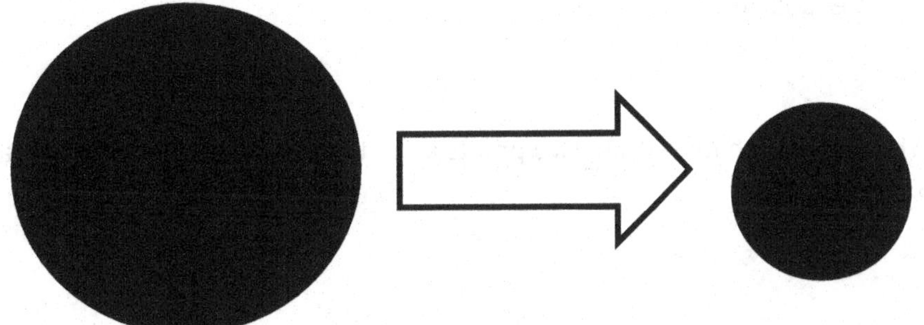

3. Follow these steps to solve Problem 1.

Step 1: $r_{new} = \frac{3}{10} r_{old}$

Step 2: $Area \; \alpha \; r^2$

Step 3: $Area_{new} \; \alpha \; r^2_{new} = \left(\frac{3}{10}\right)^2 r_{old} = \frac{9}{100}$

Step 4: $\frac{9}{100}$ = 9% of the original (old) area.

4. Follow this step to solve Problem 2.

  100%
  -9% of the original (old) area.
  ______
  91% decrease in the residual area.

APPENDIX B
Glossary

Area All that is within a circle.

Billionth billionth Prefix: nano Symbol: n Exponent: 10^{-9} Number: 0.000000001

Circumference The distance around a circle.

Constant A constant is a number which may represent a whole number or even a fraction.

Denominator The number written below the line in a fraction to let us know how many parts are in the whole to be divided.

Diameter Any straight line segment that passes through the center of the circle and whose endpoints lie on the circle.

Direct Linear Relationship When you graph more than two points in math and you end up with a straight line on the graph.

Direct Relationship A direct proportion between two mathematical variables.

Distance Equation Distance = Rate multiplied by Time. $D = R * T$

Doppler Equation $f_{DOP} = 2V\cos\Theta/c$

Exponents Shorthand for multiplication. Example: $3^2 = 3 * 3 = 9$

Flip or flipping To turn a number or fraction upside down and multiply. *See reciprocal.*

Frametime The time that it takes to make a frame, a picture.

Frame Rate The frequency at which frames are made or presented.

Frequency The number of cycles of a particular event per time.

Giga	billion Prefix: giga Symbol: G Exponent: 10^9 Number: 1,000,000,000
Hertz or Hz	A unit of frequency of a radio wave at one cycle per second.
Hundred	hundred Prefix: hecto Symbol: h Exponent: 10^2 Number: 100
Hundredth	hundreth Prefix: centi Symbol: c Exponent: 10^{-2} Number: 0.01
Indirect Relationship	An inverse relationship between two mathematical variables.
Inverse	Opposite. Example: The inverse of 3 is 1/3.
Kilo	thousand Prefix: kilo Symbol: k Exponent: 10^3 Number: 1,000
M	Mega meaning million.
Mega	million Prefix: mega Symbol: M Exponent: 10^6 Number: 1,000,000
Metric System	System of measurement based on the power of 10.
Millionth	millionth Prefix: micro Symbol: μ Exponent: 10^{-6} Number: 0.000001
Negative	A number that is less than zero and is denoted with a negative sign. Example -3.
Non-Linear Indirect Relationship	When you graph more than two points in math and you end up with a curved line on the graph.
Non-Linear Inverse Proportional	When one variable increase by X% the other variable decreases by a different percent.
Non-Linear Proportional Relationship	When one variable increases by a certain amount, the other variable increases by a HUGE amount. Example: $Y \alpha X^2$.

Numerator	The number written above the line in a fraction to let us know how many parts are in the whole.
Percent	One part in one hundred.
Period	The reciprocal of the frequency representing the amount of time it takes for one full cycle to occur.
Positive	A number that is greater than zero.
Proportional	Two variables are proportional if a change in one is always accompanied by a change in the other. Symbol: α
Pulse Repetition Frequency	The number of pulses in one second that is equal to the reciprocal of the Pulse Repetition Period.
Pulse Repetition Period	The time between the start of one transmit pulse until the start of the next one.
Radius	Any line segment from the center of a sphere or a circle to its perimeter.
Rate of Change	The ratio of change between what comes in and what goes out.
Reciprocal	A number that is multplied with its fractional counterpart that is upside down results in the number 1. For example: 3 * 1/3 = 1
Reflective Mode	Sound will reflect back from the direction it came from.
Residual Area	After decreasing a circle, how much area is left.
Scientific Notation	A method for writing very large and very small numbers using exponents.
sec	second
Ten	ten Prefix: deca Symbol: da Exponent: 10^{1} Number: 10
Tenth	tenth Prefix: deci Symbol: d Exponent: 10^{-1} Number: 0.1

Thousandth thousandth Prefix: milli Symbol: m Exponent: 10^{-3} Number: 0.001

Transmission Frequency Frequency of the wave produced to travel into the patient.

Unit A measurement.

μsec millionth of a second

Variable A variable is a quantity which can be represented by letters.

Volume How much it takes to fill up a three-dimensional object.

Index

Printed in the United States
By Bookmasters